Historic England

Ships and Boats: 1840-1950

Introductions to Heritage Assets

Summary

Historic England's Introductions to Heritage Assets (IHAs) are accessible, authoritative, illustrated summaries of what we know about specific types of archaeological site, building, landscape or marine asset. Typically they deal with subjects which lack such a summary. This can either be where the literature is dauntingly voluminous, or alternatively where little has been written. Most often it is the latter, and many IHAs bring understanding of site or building types which are neglected or little understood. Many of these are what might be thought of as 'new heritage', that is they date from after the Second World War.

This overview looks at ships and boats built after 1840. Principally drawing on archaeological, technological and historical sources, it describes vessels used on English inland and coastal waters and in the open sea. The evidence of wrecks and abandoned vessels is drawn on, as well as extant vessels. Also included is the early development of submarines.

This guidance note has been written by Mark Dunkley and edited by Paul Stamper.

It is one is of several guidance documents that can be accessed at
HistoricEngland.org.uk/listing/selection-criteria/listing-selection/ihas-buildings/

First published by English Heritage September 2012.

This edition published by Historic England July 2016.

HistoricEngland.org.uk/listing/selection-criteria/

Front cover
I K Brunel's SS Great Britain.

Contents

Introduction

Historic ships of the modern period are inextricably linked to industrial developments, and the period 1840-1950 is characterised by the progression of the marine steam engine and development of steamship technology. Paddle propulsion gradually gave way to the screw propeller, while the introduction of iron, and later steel, hulls eclipsed wooden ships and allowed the building of ever larger vessels driven by increasingly complex and powerful steam power plants. Today's concept of the standard steam engine, whereby high-pressure steam drives a piston within an enclosed cylinder, is based upon inventions that date back no further than the end of the 18th century.

The transition of Great Britain towards machine-based manufacturing from the later 18th century was made possible by the invention of the steam engine which utilised a high-pressure boiler to turn steam into mechanical action. As steam-powered machines consumed vast quantities of coal, industrialisation would never have been economic if coal could not have been readily transported by water. The Industrial Revolution drove Britain's overseas trade in the 19th century, and also facilitated the development of Britain's merchant shipping industry.

This overview addresses post-1840 vessels (understood here as being simply a general term to describe all kinds of craft irrespective of whether they were designed to navigate on the surface of the water, that is ships and boats, or under the water, that is submarines). Principally drawing from archaeological, technological and historical sources, it describes vessels used on inland waters, coastal waters and in the open sea, as well as vessels now abandoned in coastal areas.

Ship and boat remains have additional interest and significance because their construction and contents can provide important information about the social, economic and political circumstances at the time of their construction, use or loss especially when combined with documentary evidence such as technical drawings or cargo manifests.

For descriptive purposes, the remains of a vessel which has either sunk or suffered structural damage to the extent where it can no longer function, and is buried and/or submerged either in part or in whole, will be used synonymously with the term *wreck*.

The number of designated vessels is small but with some 37,000 known wreck sites and recorded ship losses in the English Territorial Sea (which extends up to 12 nautical miles from the English coast), there is a recognised need for pragmatism in increasing the number of designated sites. This introduction, which in terms of designation interest should be read alongside the selection guide on **Ships and Boats** and complements the **Ships and Boats: Prehistory to 1840** Introduction to Heritage Assets, which describes vessels from about 1840 to 1950. This was a period of major change in shipbuilding when the full impact of the industrial revolution and Britain's commercial and colonial expansion was felt in the maritime world, particularly in relation to the *Pax Britannica* (the period of relative peace in Europe and the world between 1815 and 1914), when in little more than a hundred years sail had given way to steam and nuclear propulsion and wooden hulls to those made of iron and then steel. The known historic maritime resource in waters off England is dominated by wrecks of the mid 19th to mid 20th centuries: 96 per cent of known and dated wrecks were lost in the period between 1840 and 1950.

As vessels and shipwrecks in archaeological contexts rarely survive in their entirety, coupled with other vessels being broken-up at the end of their working life, both museum exhibits and those vessels forming the National Historic Fleet within the National Register of Historic Vessels (being those vessels of pre-eminent national or regional significance, administered by National Historic Ships UK) and the National Small Boat Register (administered by National Maritime Museum Cornwall) are also included in this narrative. It is notable that ships and boats in preservation built between 1860 and 1913 outnumber those recorded in the known archaeological resource.

1 Description and Chronology

This is a rapid introductory survey of the range and chronology of post-1840 ships, boats and submarines concentrating on the larger types of ship. It identifies significant vessel types, and notes where there are gaps in our understanding: it is not a comprehensive review of known vessel types, museum exhibits, marine engines, wreck sites or working boats, nor is it a synthesis of 19th and early to mid 20th century British maritime history, for which Friel (2003) and Griffiths (2001) are recommended.

1.1 Victorian (1837-1901)

On 19 July 1837, a month after Queen Victoria acceded to the throne, Isambard Kingdom Brunel's revolutionary steamship, the *Great Western*, was launched at Bristol. Built specifically for passenger transport across the Atlantic, the *Great Western*'s wooden hull design was based on those used in the line-of-battle ships of the day, but with emphasis on longitudinal strength (that is, from end to end) to resist the buffeting of the Atlantic waves. The engines, built at Lambeth, were the largest built up to that time, and drew on steam from four iron boilers, each having three furnaces (though the presence of secondary masts and sails afforded a measure of emergency propulsion).

Brunel's designs were based on nearly fifty years of experimentation. In 1802, the *Charlotte Dundas* (considered to be the first practical steamboat) towed two 70-ton barges along part of the Forth & Clyde Canal, and by 1812 advances in boiler and cylinder design had enabled the Clyde-based paddle steamer (PS) *Comet* to provide the first commercially successful steamboat service (Figure 1) and a great number of boats of this kind were built to work in river estuaries. Their use was limited for they were only designed to work on smooth water as well as being constrained by their high fuel consumption.

The advantage of steam was that it provided a source of power independent of the wind and so it was easier for steamships to keep to a regular schedule. From 1815 the PS *Margery* was providing a ferry service along the Thames with the first regular cross-Channel service by steam following in 1821. As crossing times were reduced to 2¾ hours (as opposed to at least three to four hours under sail), the advantage of steam power for short sea journeys was becoming indisputable, and during the 1820s new shipyards in London were building steamboats not only for the Channel service, but for other routes along the English coasts.

Although in the course of the 1830s steam services were beginning to extend across the Atlantic and to the Far East, there were inherent difficulties associated with the supply of fuel: the historian Thomas Crump, for example, records

Figure 1
Launched in 1812 at Port Glasgow (and wrecked in 1820), the *Comet* was the first commercial steamship in Europe. A replica stands in Port Glasgow town centre. © clydesite.co.uk

that the *Bernice* spent twenty-five days in harbour coaling out of a voyage of eighty-eight days from Falmouth to Calcutta. What is more, until the widespread introduction of surface condensers (which allowed ships to recycle and steam indefinitely on fresh water, which also permitted a saving in fuel consumption) from 1834, boilers had to be regularly descaled owing to the reliance on sea water to provide the necessary steam.

Brunel realised that a substantial increase in ship size would be necessary if it was to carry sufficient coal to cross the Atlantic. By determining the correct ratios, it was possible to calculate the minimum dimensions of a steamship able to carry sufficient fuel for a specified distance. The Great Western Steamship Company was formed in 1836 to build a line of steamships for the Bristol-New York route with the hull of the *Great Western* being laid down in June of that year. The mass-migration these facilitated was to play a large part in shaping the American nation.

In the 25 years that passed between the *Comet* and the launching of the *Great Western*, and despite improvements in steamship construction, vessels belonging to this first generation had certain fundamental common features. Apart from a few notable exceptions such as the iron-hulled *Aaron Manby* (trialled on the Thames in 1822; broken up in 1855), their hulls were built of wood; they were driven by paddle wheels housed in paddle boxes amidships; and their single-expansion engines (where steam is expanded through only one stage causing all cylinders to operate at the same pressure), also located amidships, exhausted steam into jet condensers (condensing the exhaust steam at the end of a power stroke made a steam engine more efficient) which were basically similar to the engine patented by James Watt in 1769. Sails continued to be carried as a safety measure.

The first steam-driven fighting ship of the Royal Navy, HMS *Comet*, was launched in 1822 at

Deptford, and although paddle frigates were introduced in the early 1840s there was no fundamental change in naval policy. The historian K. T. Rowland observed that one of the chief objections to the use of steam in warships was that the engines were considered vulnerable and could not be protected by being placed below the waterline because they were too bulky. Although the use of steam propulsion freed warships from their dependence upon the wind, early steam-driven naval vessels were simply regarded as useful auxiliaries (to be used as tugs when there was no wind) whose role was supplementary to the wooden line-of-battle ships. The navy was still committed to completing the *Vanguard* class of Second Rate wooden warships among others, launched between August 1835 and July 1848.

It therefore seemed unlikely that steam could ever compete economically with sail, particularly for carrying long-distance cargoes, and this coupled with an inherent resistance against the adoption of iron for ships' hulls represented a radical departure from traditional building techniques. However, the production of wrought iron in Britain was lowering the price of iron plate, while timber was becoming scarcer and dearer. As Britain had to import most of its shipbuilding timber by this time, economic considerations slowly began to favour the construction of iron steamboats.

Although iron ships were still comparatively rare by the end of the 1830s, the construction of rolling mills able to manufacture large iron plates, such as the scheduled complex at Little Matlock, Sheffield, made the possibility of building an ocean-going ship in iron practicable. The directors of the Great Western Steamship Company therefore agreed tobuild another ship, larger and faster than the *Great Western* (which was to be scrapped in 1856 after serving as a troopship during the Crimean War), and the keel of the SS *Great Britain* was laid down in 1839. The *Great Britain* (which is now on display in Bristol and part of the National Historic Fleet; see also front cover illustration) was to embody two important developments that were eventually universally adopted by all naval architects: the use of iron as a shipbuilding material and the incorporation of a screw propeller driven, in this case by a direct-acting engine which applied power directly to the crankshaft.

Although slow to recognise the potential of steam power for the battle fleet, by 1837 the Admiralty had 37 steamboats, comprising tugs, dredgers and packets and had established a special steam department. It was never the intention of the Admiralty that steam power should be used exclusively as designs for lifting screws and lowering funnels were adopted particularly as engines were only generally engaged for moving in and out of harbour and manoeuvring in confined waters.

By 1845, experimentation concerning the relative merits of screw and paddle propulsion culminated in the celebrated towing contest between the screw-driven HMS *Rattler* and the paddle frigate

Figure 2
Figurehead of the *Thames*, built in 1827 and wrecked in 1841, in Tresco Abbey Gardens, Isles of Scilly. The remains of the *Thames* are important as it was a very early paddle steamer. Ship relics such as this are often of significance, and many are now listed.
© Alasdair Moore, Tresco Abbey Gardens Valhalla

HMS *Alecto*. During the trials, *Rattler* towed *Alecto* astern at a speed of 2.8 knots which signified the beginning of the end of paddle-driven vessels in the Royal Navy.

The main changeover from paddle to screw propulsion occurred gradually over 20 years from the 1840s. The Cunard Company, for example, continued to order paddle steamers until the early 1860s. Only two vessels are known from this period: the *Thames* (wrecked 1841, Isles of Scilly) and *Pegasus* (wrecked 1843, off Holy Island). The figurehead of the Dublin packet *Thames* was recovered and can now be seen in the Tresco Abbey Gardens, Isles of Scilly (Figure 2). An exceptional late employment of paddle-propulsion was in the Second World War, for purpose-built minesweepers.

Despite the advantages offered by steam technology, it was not immediately adopted universally, and wooden brigs and snows (large, two-masted, merchant ships) remained important trading vessels until the early 20th century. The wreck of a snow, the *Douro* (wrecked 1843 off the Isles of Scilly), is the only known wreck site of a sailing ship lost during the 1840s. From the large number of bronze manillas (slave tokens) found on this wreck site there is speculation that the Sunderland-registered vessel was involved in the slave trade, and was onward-bound to Africa despite the Slavery Abolition Act of 1833.

Britain's great supplies of iron ore and coal provided enormous advantages during the Industrial Revolution, enabling the country to produce many metal items, including engines, more cheaply than its competitors. A wide variety of marine steam engines were developed over the course of the 19th century classified by either their connection mechanism (the means of supplying power to a crankshaft), such as side-lever, direct acting, oscillating or vertical, or cylinder technology, such as simple expansion, compound or triple expansion. As most early marine engines had the same cylinder technology (where all cylinders operate at the same pressure) they are generally classified according to their connection mechanism. One early survivor of a single-cylinder engine is the drag-boat *Bertha*, built in 1844 and supplied to the Port of Bridgwater (Figure 3). *Bertha* is now the oldest operational steam vessel in Britain and forms part of the National Historic Fleet. Throughout the 19th century, as most cylinder technologies were growing more complex, engines began to be classified solely according to their cylinder technology (but inevitably one can find examples in the literature of engines which were classified under both methods!).

Figure 3
The chain-operated drag-boat *Bertha* was built in 1844 to act as a dredger for the inland port of Bridgwater. It is the oldest operational steam vessel in Britain, and now forms part of the International Sailing Craft Association collection and National Historic Fleet. She is on display at Eyemouth (Berwickshire).
© Eyemouth International Sailing Craft Association

The use of iron for shipbuilding, the adoption of the screw propeller and, from 1834, the introduction of surface condensers, were significant stages in the development of the steamship. These developments influenced the thinking of any naval architect called upon to design a typical steamship in the period between 1850 and 1860, like the *Faith*. Constructed in Birkenhead in 1852, this iron-hulled vessel was propelled by both steam and sail, and had served as a Crimean War troopship before foundering off the Isle of Wight three years later *en route* to Istanbul to be delivered to new Turkish owners. Only two other iron steamers from this period are known: the wrecks *Nile* and *Zebra*.

The Crimean War (October 1853 to February 1856), fought to gain influence over the territories of the declining Ottoman Empire, is considered to be one of the first modern wars because of the tactical use of new technology like the electric telegraph and railways. Although the Allied (Britain, France, Turkey and the Kingdom of Sardinia) naval campaign was a success, the war showed that pure wooden sailing ships were particularly vulnerable to accurate fire from modern shore-based artillery; the French were forced to protect their ships with iron armour. This idea was copied by the Royal Navy and resulted in the *Aetna*-class iron-armoured floating batteries of 1855/6: most of the Navy's sailing ships were then converted.

The final step in the development of the 'ironclads' was taken by the Admiralty which ordered the building of the *Warrior* and *Black Prince* in 1859. Built at the Limmo Peninsula shipyard at Blackwall (London) in 1860, HMS *Warrior* survives as part of the National Historic Fleet in Portsmouth (Figure 4), while HMS *Black Prince* was hulked in 1896 and sold for scrap in 1923. These were the first large warships designed with iron hulls and protected with iron armour. When they were built, they were the most powerful fighting ships in the world and changed the balance of naval supremacy – but rapid advances in naval technology and tactics, especially during the American Civil War (when a submarine was first deployed successfully), left these two *Warrior*-class ironclads obsolete within a short time.

Figure 4
Built at Blackwall in 1860, HMS *Warrior*, one of the first ironclads, is part of the National Historic Fleet in Portsmouth.

A further example of just how far naval attitudes had changed from the 1840s is exemplified by the conversion of the 120-gun First Rate Ship of the Line HMS *Royal William* (later HMS *Clarence*), launched in 1833, to a 72-gun screw ship in 1860. The *Clarence*, which was accidentally burnt in the River Mersey in 1899, comprises the only known wreck of a warship known to have been converted from sail to screw propulsion.

However, many of the critical advances in naval engineering were pioneered by the mercantile community, where potential profit provided a fierce driver for progress. Aware of the tremendous increase in trade which had occurred between Britain and Australia, Brunel proposed to build a vessel large enough to sail to Australia and back without coaling. A bunker capacity of 10,000 tons was required. Limitations were imposed on the ship's draught and the resulting ship, the *Leviathan* (as the *Great Eastern* was originally named) was laid down in May 1854 at Millwall on the Thames and launched three years later (Figure 5). The *Great Eastern* was to be the only steamship ever built powered by both screw and paddles. Although equipped to carry 4,000 passengers, Brunel had miscalculated the demand for such a ship. The *Great Eastern* was therefore later converted to a cable-laying ship and subsequently a floating music hall before being broken up in 1888/9 (a funnel from the vessel is on display at the SS *Great Britain* museum, Bristol).

Nevertheless, the merchant marine of the north-east continued to satisfy the demands for coal in

Figure 5
Isambard Kingdom Brunel's steamship SS *Great Eastern* under construction at Millwall, London, in 1857. The largest ship ever built at the time of her launch, she had a chequered career before being broken up in 1889-90.

the capital. In 1844 2.5 million tons of coal was shipped to London in 9,500 separate voyages, and every year from 1851 until the turn of the century the volume of coal shipped increased. Vessels like the wooden collier *Rising Sun*, wrecked off Garrison Point, Cleveland, in 1860, were already in decline and being replaced by iron screw colliers. In 1904 the last collier brig, the *Remembrance*, sank in a gale off Aldeburgh in Suffolk.

The change to iron ships produced a shift in the location of the shipbuilding industry which had grown up mainly in the south of England and the Thames. While timber for shipbuilding was available in the south (whether home-grown or imported), the raw materials and the necessary mechanical skills to build metal-hulled ships were in the north. The industry moved to where there was coal, iron and sheltered waters: Clydeside, Merseyside, Tyneside and the coast of Durham.

The volume of shipping built at this time is staggering. In 1860, for example, 178 vessels were built on the Clyde alone including the paddle steamer *Iona II*. Built in 1863 as a fast ferry to ply the Clyde, the *Iona II* was exceptionally well fitted out. The specially designed twin-cylinder engines were oscillating (where the cylinders pivoted as the crankshaft rotated, thus providing a reduction in engine size and weight) and fitted with tubular boilers, superheaters and every well-tried improvement. The steamer had luxury passenger accommodation, a 75ft dining room and 180ft saloon with velvet sofas and reputedly gave a top speed of 24 knots. Given this speed, the *Iona II* was soon acquired as a blockade runner to run guns and supplies to the Confederate forces in the American Civil War. It is probable that she was running without lights in dense fog to avoid detection when she foundered east of Lundy Island, North Devon on her first transatlantic voyage in 1864. Following her discovery in 1976, the *Iona II* was designated a Protected Wreck Site in 1989 (Figure 6).

The fishing fleet was also embracing the new technologies, and in 1877 a shipbuilder in North Shields announced the first paddle-steamer boat to pull a beam trawl. The sail-powered smack, however, remained the most numerous east coast fishing vessel of the 19th century covering the deep-water grounds from Greenland to the shoals of the Dogger Bank. In the 30 years from 1845, the North Sea fleet had grown from 29 smacks to around 400. By 1880 there was also a fleet of paddle-trawlers operating out of Scarborough, yet by 1904 they had all been replaced by propeller-driven trawlers, an industry centred on Grimsby and Hull; several such trawlers are recorded as wreck sites around England, such as the *Dayrian*, built in 1906. Just a handful of smacks, such as the *Ellen* (built in 1900 and now a Registered Historic Ship) survive but only one is recorded as a wreck: the *Gwydir*, wrecked in 1902 off Liverpool. Similarly only one herring drifter (albeit a steam boat) survives from the east coast herring fleets: the *Lydia Eva* was built in 1930 and now forms part of the National Historic Fleet. Though the backbone of the Cornish fishing fleet was the lugger (so called because of its lugsail; a four-sided sail set on a lug, or yard), the only known survivors are a handful of vessels, such as the *Ripple* (built in St Ives in 1896), all Registered as Historic Ships. Brixham sailing trawlers continued to be built and used until the 20th century; a number survive.

By 1860, however, steam was still playing a secondary role as far as long-distance bulk cargo carrying was concerned, and vessels equipped with single-expansion engines were no match for the legendary clipper ships (which, like most sailing vessels, had no fuel bills, smaller crews and no requirement for machinery repairs). These vessels, like the Grade-I Listed *Cutty Sark* (built 1869 and now part of the National Historic Fleet), were developed specially for transporting expensive cargoes such as tea and wool half-way round the world from China and Australia to the London markets (Figure 7). Further, a series of wooden ‘Blackwall frigates’ were built between 1837 and 1869, following the expiry of the East India Company’s monopoly on eastern trade in 1833. These large, fast, three-masted sailing ships were employed on Indian trade via the Cape of Good Hope, though none are known to survive. They do, however, indicate the continued investment in building wooden sailing ships during the mid 19th century.

Figure 6
Wreck of the designated Paddle Steamer *Iona II* (launched 1864), off Lundy, Bristol Channel, showing paddle wheel debris.

Figure 7
A detailed view (taken c.1960), of the bow and rigging of the *Cutty Sark* in its dry dock at Greenwich, looking up from the ground towards the foremast.

The 1860s were the zenith of the merchant sailing ship era, for while 16 British and American Clippers competed in the 1866 tea race to bring the annual tea crop from Foochow (now Fuzhou), China, to London, the opening of the Suez Canal in 1869 hastened the decline of the sailing vessel for long-distance voyages. The Suez Canal shortened the route to India and the Far East by several thousand miles and steamships gradually came to dominate the lucrative Indian trade as sailing ships were not adaptable for use through the Canal (as the prevailing winds of the Mediterranean blow from west to east) and when homeward-bound had to make the journey round the Cape of Good Hope.

However, by far the most significant development within marine engineering that impacted directly on sailing ships was the widespread adoption of the compound engine. These engines (where steam is expanded in two, or more, stages), were strong enough to contain superheated steam at high pressure, and had been employed in Lancashire textile mills from the 1850s where they could claim fuel savings of up to 40 per cent when compared to a single-expansion engine of the same power. The economy in fuel (and associated increase in cargo capacity) attracted ship owners to this type of engine for use in long-haul cargo and passenger traffic, though it was only in the 1860s that such engines became sufficiently reliable for use in ocean-going vessels. Compound engines were later adopted for practically all ships built for the transatlantic service, and during the five years between 1870 and 1875 the number of ships fitted with such engines trebled until the total exceeded over two thousand vessels.

Although the iron schooner *Ida*, built in 1849 and wrecked 1871 off Sefton, Merseyside, comprises a very early example of a steamship fitted with a compound engine, the majority of known wreck sites from this period with compound engines are the remains of transatlantic vessels, which at this time was diversifying into specialist routes and traffic. Most shipping countries had 'liners', that is passenger and freight ships sailing by fixed routes at regular and advertised dates. While the emergence of shipping companies carrying passengers on scheduled routes dates from the 1840s, this business was to fall into the hands of large shipping companies like the Cunard and White Star Companies (which later merged), the P&O Company on the eastern routes and Union Castle on the Africa run (Figure 8). Wrecks such as the *Deutschland* characterise the early types of passenger ship. Sailing from Bremerhaven with 123 emigrants bound for New York via Southampton in December 1875, the *Deutschland* ran aground off Harwich in a blizzard: only 135 persons were saved out of 213 passengers and crew, inspiring Gerard Manley Hopkins' poem *The Wreck of the Deutschland*. Liners, such as the SS *Schiller* (built 1873 and wrecked off the Isles of Scilly in 1875) carried passengers only, while cargo liners, such as Cunard's SS *Sromboli* (built 1856 and wrecked off the Lizard in 1878) were cargo-carrying vessels with accommodation for a few passengers.

In addition, a new class of merchant vessels developed that eventually displaced the last generation of sailing ships. Long-haul cargo tramp steamers were so-called because rather than working a regular route they carried general cargo to any destination as required, directed by the owners by telegraph and later by radio. Other ships began to be designed for particular purposes. From the 1880s oil tankers were built for carrying oil in bulk from the oilfields of the world to the countries which began increasingly to use it. Other new types included coasters, grain ships, refrigeration ships, cattle ships, whale factory ships and cable-laying ships. In most of these developments Britain played a leading part, both in construction and ownership, managing to hold the lead into the 20th century.

One survivor of the diversification of shipping is the coaster SS *Robin*. Built at the Thames Ironworks, Orchard House Yard in East London and launched in 1890, the *Robin* is the only

Figure 8
SS *Britannic* operated for the White Star Line between 1874 and 1902, before being scrapped in 1903. A steamship equipped with sails, she carried 266 Saloon (that is, First) Class passengers, and 1,500 steerage passengers.
© From the collection of Björn Larsson

complete example in the world of a coastal cargo steamer and is also the world's oldest complete steamship. The *Robin* now forms part of the National Historic Fleet and is temporarily moored at London's Royal Victoria Dock.

The development of such steamships put pressure on existing dock systems, many of them too small to accommodate the larger iron and steel ships. Although the modern harbour at Dover was opened as late as 1909, the first stones of Southampton's modern docks were laid in 1838. That facility opened in 1843 and was heavily used by the P&O Line which had secured a contract to run mail to India. Unsurprisingly, Southampton, rather than London, became the major transoceanic passenger port, and was also the principal embarkation port for men and material during the Boer Wars (1880-1881 and 1899-1902). Both the White Star Line (operators of the future *Titanic*, the biggest ocean steamer ever built) and the American Star Line relocated their transatlantic termini from Liverpool to Southampton in 1895.

Britain also embarked on the process of raising shipping standards in law, led by Samuel Plimsoll (MP for Derby), beginning with the Merchant Shipping Act of 1876 which made the marking of a Load Line compulsory. Ships were not to be loaded so that the line was submerged. This was the so-called Plimsoll Line, modified to allow for loading in different types of water. Subsequent legislation provided for better standards for crews, and the Board of Trade established a qualification scheme for officers.

Saving lives at sea around the coast of Great Britain by a national lifeboat service began in 1824 with the founding of the National Institution for the Preservation of Life from Shipwreck (later the Royal National Lifeboat Institution or RNLI). The Institution was able to create a chain of lifeboat stations manned by volunteer sailors and fisherman and their courage is well known. Lifeboats like the Norfolk and Suffolk Class *Alfred Corry*, built in 1893 in Great Yarmouth (and now part of the National Historic Fleet on display in Southwold) and the listed *Tyne* of 1833 at South Shields, link 'terrestrial' assets like the Grade-II listed Lifeboat House and Slipway at Birnbeck Pier, North Somerset, with their maritime rescue function. All lifeboats were driven by sail and oar until the late 1890s when a few steam lifeboats were built. Some rescue independent services remain alongside the RNLI, such as the South Shields Volunteer Life Brigade. For lifeboat buildings see the **Maritime and Naval Buildings** selection guide. See too **https://www.historicengland.org.uk/images-books/publications/iha-coastguard-stations/**

The Admiralty was to show a keen interest in improvements in thermal efficiency brought by the compound engine, mainly because of the diminution of coal stocks in overseas coaling stations. An experimental programme undertaken during the 1860s saw the conversion of several wooden corvettes, such as HMS *Clarence*, to trial new engines. By the mid 1870s, the Admiralty was trialling compound engines in the larger *Iris* class cruisers HMS *Iris* and *Mercury* launched in 1877 and 1878. Despite being steam ships driven by twin screws, both cruisers were equipped with a light barque sailing rig on completion and were also the Navy's first all-steel ships.

Changes in steel production in the second half of the 19th century enabled the development of an inexpensive method for its mass-production from molten pig iron. As in warships, steel began to replace iron in merchant shipbuilding from the 1880s onwards following a *Report on Steel for Shipbuilding Purposes* prepared for Lloyd's Register of Shipping during 1877-8. At the same time, 1876 was the first year in which the number of steamships built exceeded those under sail.

The reduction in scantlings (the dimensions of all parts of a hull) provided by steel construction, combined with the Royal Navy's adoption of self-propelled torpedoes in the 1870s, gave rise to new specialised light craft such as the fast 'torpedo boat' of which the wreck of HMS *Landrail* is the only known example of this period. The steel-hulled *Landrail* was built in 1886 and had the honour of being one of the first ships to pass through the newly opened Tower Bridge in June 1894; she sank while under tow in Lyme Bay in

Figure 9
HMS *Gannet*, launched 1878. This class of warship undertook trade protection, anti-slavery and surveying duties. The Gannet was preserved in 1987 and is now part of the Core Collection of the UK's National Historic Fleet. © Chatham Historic Dockyard Trust

1906. To counter torpedo threats from foreign navies, the 'torpedo-boat destroyer' (later to be simply called 'destroyer') was developed in the 1880s, and became a major new type of escort vessel. The only known wreck of an early torpedo-boat destroyer is the steel-hulled HMS *Decoy* which foundered in a collision off the Isles of Scilly in 1904 during night exercises. Another new type of craft was the submarine which was becoming a practical weapon by the latter half of the 19th century (see separate section below).

Other innovations were to follow throughout the 1870s: the first steamship to transport oil was built in 1872 in Tyneside, auxiliary machinery to drive steering gear, bilge pumps, feed pumps, evaporators, capstans and, in warships, gun turrets was introduced, and in 1876, two British warships, HMS *Minotaur* and HMS *Temeraire*, were equipped with steam-driven generators for electric arc searchlights. Warships incorporated the latest technological developments as they became available, and electric arc searchlights were soon adopted in the fitting out of all large warships.

Eventually the use of sail was also recognised as seriously impairing the fighting efficiency of the new designs. As a result, the *Devastation* was commissioned in 1873 as the first of a class of ocean-going capital ship that did not carry sails; this has since been described as the most radical design to emerge in the 19th century, marking the beginning of end of the era of the sailing warship. However, the Navy was not yet fully ready to break with tradition for 1878 saw the completion of HMS *Gannet*, which is now the last surviving archetypal Victorian 'Gunboat' and part of the National Historic Fleet. Classed as a screw sloop and thus a clear example of the transition from sail to steam, the *Gannet* was of composite construction (with a wooden hull over an iron frame) fitted with a compound engine; she was deployed overseas to police the trade routes that served the British Empire until 1895. The *Gannet* was converted into a drill ship in 1902 and is now on display in Chatham Historic Dockyard and maintained by the Historic Warships Volunteer Group (Figure 9).

Figure 10
Radial flow turbine engine of the Turbinia, launched 1894, the world's first steam turbine driven vessel. On display in the Marine Engineering Gallery, the Science Museum, London
© Mark Dunkley

The success of the compound engine was followed in 1874 by the even-more powerful and efficient triple-expansion engine. Here, a third intermediate cylinder was added to the high and low pressure cylinders of the compound engine necessitating a boiler capable of withstanding even higher pressures. The saving in coal made larger ships possible, signalling the beginning of the modern transatlantic liner. The British-built *City of Paris* – the first ship to make the crossing in under six days – and the *City of New York* entered service in 1888 (but were scrapped in 1923 and 1922 respectively). Wreck sites such as the German *Patria* (foundered in the Downs, off Kent, in 1899) and the steam yacht *Argonaut* (lost off Rye, 1908, prompting a Parliamentary Question regarding lifeboat accommodation and a subsequent Public Inquiry) offer the opportunity to investigate such craft.

Ordered in 1889, HMS *Hood* was the last of eight *Royal Sovereign*-class pre-dreadnought battleships built for the Royal Navy in the 1890s. Fitted with coal-fired triple expansion engines with improvements in firepower, armour and speed, the Royal Sovereigns were the most powerful battleships in the world (until rendered obsolete by the *Dreadnought* in 1906). The *Hood* was decommissioned in 1911 and was later scuttled to act as a blockship across the southern entrance of Portland Harbour (Dorset) at the beginning of the First World War.

The first quadruple expansion marine engine was fitted into the *County of York* which was built in Barrow in 1884 (later to be renamed and run aground off Nexø, Denmark, in 1906). For a comparatively short time this type of engine was the most powerful engine afloat but the technological limit had been reached: piston-driven engines were already being supplanted by the steam turbine.

The British engineer Charles Parsons revolutionized marine propulsion in the 1890s with the development of his steam turbine, which had a very much higher power to size/weight ratio than any preceding engine type. Parsons (1854-1931) developed a radial flow for steam, where steam is introduced near the shaft and made to flow radially outwards before being conducted back inwards towards the shaft.

Following successful installation of steam turbines to generate electric power for the Newcastle and District Electric Light Company in 1888, Parsons founded the Marine Steam Turbine Company to produce turbines suitable for maritime use. The *Turbinia*, built in 1894 (and now part of the National Historic Fleet with the hull on display in the Newcastle upon Tyne Discovery museum), became the world's first steam turbine-driven vessel, and sea trials in 1897 saw her achieve maximum speeds of 31 knots (compared to 17.15 knots achieved by the *Cutty Sark*). The engine from *Turbinia* is on display in the Science Museum, London (Figure 10).

Though the Institution of Naval Architects was less than enthusiastic about *Turbinia*, the Admiralty was interested in the new technology given the naval arms race with France and Germany. Two new destroyers, HMS *Cobra* and HMS *Viper*, were ready for sea trials in 1899. Though their performance exceeded expectation, the ships were lost in 1901 after foundering on rocks off Cromer and Alderney in bad weather. However, the success of the trials saw the introduction of turbine-driven merchant ships in the same year.

Before the competition from steam-driven vessels became too intense towards the end of the 19th century, schooners were employed in regular trade around the coastline of Britain, Europe and across the Atlantic, carrying cargoes from port to port throughout the 1860s and onwards into the 20th century. Schooners were merchant ships with two or more masts with sails aligned fore-and-aft; the crucial decade for these vessels was 1870-1880 when they had established themselves as economical and efficient vessels for all trades at home and on deep water. Indeed, the period from 1870 until the First World War has been called the era of the merchant schooner.

Schooners like the wrecked wooden *Fenna* (built 1863) and the iron *Alster* (built 1867 and recorded as being mainly intact, off north Norfolk) coupled with the schooner-rigged *Violette* (built 1919 and now part of the National Historic Fleet as a very early example of ferro-cement build) are indicative of this period from about 1860 when iron was increasingly being preferred to wood in construction and steam replacing sail for propulsion.

For inland waterways, the great age of British canals – generally accepted as being the last quarter of the 18th century – came a generation too early for any account to be taken of the advantages of steam power for boats. Canals were too narrow for navigation by boats carrying the weight of a steam engine. However, the demand for coal produced by mines close to rivers saw the development of steamboats designed to ply Britain's waterways – both natural and artificial – to reach distant consumers as distant as the steam-driven pumps of the Cornish tin mines. Vessels like the Humber Keel *Pioneer*, now a Registered Historic Ship, are representative of the type of vessel employed in this work.

Another link to Cornish industry is the remains of a mid-19th century wreck off Little Ganinick, Isles of Scilly. Here, a discrete cargo mound consisting of components of mining equipment was discovered in 2005 by local divers. The majority of the components appear to have been intended for use as pumping equipment and it is likely that this cargo represents a consignment from a Cornish foundry. The worldwide expansion of the mining industry in the 19th century was one of the most significant periods in the history of Cornwall, and resulted in massive migration and the spread of Cornish culture throughout the world. As the inscription of the Cornish mining landscapes as a World Heritage Site lends even greater weight to any surviving unaltered evidence of mining machinery from this period, the wreck was Registered in 2007.

Traditional wooden barges, like the scheduled *Harriett* at Purton, (Gloucestershire) and the Registered *Advance*, continued to be built as were spritsail barges like the *May*. Built in 1891 in Harwich, the *May* was employed to transport grain between the London Docks and flour mills in Ipswich and now forms part of the National Historic Fleet. At the

Figure 11
Composed of material retained from the RMS *Mauretania*, the interior of the Grade II-listed Mauretania Bar & Lounge, in Bristol, shows the luxurious accommodation afforded by the early transatlantic liners.
© Mauretania Bar & Lounge, Bristol

same time, smaller trading vessels like the 1898 wherry *Albion* were gradually being replaced by motorised wherries like the *Jester*. Both wherries are Registered Historic Vessels.

The increase in vessels used for recreational activities, and the rise of sailing as a hobby and then a sport (with attendant facilities like club houses, as at Cowes, on the Isle of Wight: for these see the **Sports and Recreation Buildings** listing selection guide) is evidenced by the number of pleasure craft in the National Register of Historic Vessels. These include pleasure boats (eg *Beatrice*, 1890), yachts (eg *Carola*, 1898), saloon launches (eg *Donola*, 1893) and passenger boats (eg *Gaiety*, 1889), as well as those on the National Small Boat Register (eg steam launch *Birdie*, 1899) and those curated as part of the Eyemouth International Sailing Craft Association (eg a 14ft canoe, 1860). The Registered narrowboat *Holland*, 1899, was converted to a passenger launch with a capacity of 49 passengers and operates daily cruises on the Regent's Canal between Little Venice and Camden Lock.

1.2 Early Twentieth Century (1901-1932)

The first turbine-driven liners, the British-owned Allan Line *Virginian* and *Victorian*, had entered service by 1904 and by 1907 sea trials of the Royal Mail Ship *Mauretania* (scrapped 1934/5; Figure 11) and RMS *Lusitania* (torpedoed 1915) ushered in the great age, in terms of speed, size and comfort, of ocean travel.

However, despite the progress made in marine power-plants, the grandest type of merchant sailing ships were still rolling off slipways. These 'windjammers' were the ultimate type of large sailing ship having a steel hull and between three and five masts. Such cargo ships were designed for circumnavigation. Several windjammers are still in existence (such as the Russian *Sedov*, built as the *Magdalene Vinnen II* in 1921), while the remains of the German *Preussen* are visible at low spring tides in Crab Bay, Dover, where she foundered in November 1910.

That said, a major transformation in the fuel burnt by powered ocean-going ships was already taking place. From the 1870s, the development of natural oil resources on the Caspian Sea provided an alternative to coal because of its higher calorific value which enabled vessels to attain greater ranges and higher speeds than could be obtained for the same weight of coal. In addition, the time required for bunkering was reduced and employment savings could be made by reducing the number of stokers required. In 1881 a British yard fitted the SS *Gretzia* to burn oil, and oil-powered ships for the Royal Navy followed from 1900. Although work to replace coaling stations began almost immediately, British destroyers built under the 1908-9 programme still burnt coal. Only by 1911 were oil-bunkering facilities sufficient; the new *Acheron*-class destroyers laid down in that year all burnt oil. Some existing merchant ships were converted to the new fuel.

Another innovation in the first decade of the 20th century also had consequences for world shipping. In 1898 the Italian inventor, Guglielmo Marconi, succeeded in transmitting radio signals across the English Channel. Ship-to-shore communications followed in 1901 when the British ship, the *Lake Champlain*, made a wireless link to a shore station on the Isle of Wight in May 1901. By the end of the year, there were eight commercial stations in Britain. Marconi operators were able to summon help when two liners collided off Massachusetts in 1909, while Jack Phillips and Harold Bride stayed at their posts until the last moment transmitting distress calls in the North Atlantic following the collision of the RMS *Titanic* with an iceberg in April 1912. In the aftermath of the disaster, the main focus was on the regulations governing safety of ships at sea with the first of the International Conventions for the Safety of Life at Sea being held in 1914.

The *Titanic* was one of many liners transporting emigrants (and others) to North America, and the volume of required shipping is indicated by the level of immigration; in 1907 for example, 1,004,756 arrivals were counted entering the United States. During this great age of steam that ended with the First World War Britain retained 61 per cent of the world's shipbuilding market, while the absence of any rival form of transport generated unprecedented long-distance passenger transport and imports to be delivered. New technologies, such as refrigerated transport, affected diets at home through the lowering of the price of meat.

The naval armaments race leading up to the First World War led to both Britain and Germany (as well as the US, France, Italy and Japan) building warships of previously unknown size and power. For the British, the race culminated in the *Dreadnought*, launched in February 1906, which was the first large warship to be turbine driven. Very heavily armed and powered, she rendered all earlier battleships obsolete. Innovation continued, and in 1912 HMS *Bristol* became the first warship to run on superheated steam from her twelve boilers, enabling even greater speeds as well as fuel economies.

Although there was some general progress in marine engineering during the years of the First World War, the Admiralty pursued a policy of caution and concentrated on simplifying machinery layout. As the war progressed, the main problem was to keep ships at sea or in a state of constant readiness. Maintenance had to be reduced to a minimum and rapid repair through the use of small-scale welding was introduced from 1917 as an alternative to the standard practice of riveting hull plates together.

In February 1915, however, Germany declared the waters around the British Isles to be a war zone, with merchant ships – Allied and neutral alike – subject to attack. By September 1915, U-boats had sank 480 merchant vessels in British waters including the Cunard liner *Lusitania*, which was torpedoed off the Irish coast in May 1915 with the loss of 1,201 men, women and children. Anti-submarine measures employed by the Royal Navy initially relied on crude weapons, patrols and Q-ships (armed ships disguised as merchant or fishing vessels). Convoy systems were introduced from April 1916 as the Allies were losing an average of 65 merchant ships for every U-boat sunk (a figure which rose to 167 in April 1917).

The largest naval battle of the war was fought on 31 May and 1 June 1916, some distance off the Danish Jutland coast. The aim of the German fleet was to bring the British Grand Fleet to battle and to defeat it. The battle cost the lives of over 8,000 sailors and sank 25 ships (Figure 12). In May 2006 on the ninetieth anniversary of the battle, the remains of the fourteen British ships lost there were designated under the Protection of Military Remains Act 1986. The light cruiser HMS *Caroline*, the sole survivor of the battle and a National Historic Fleet vessel moored in Belfast, has been gifted to the National Museum of the Royal Navy.

Advances in marine engineering which had occurred since the turn of the century influenced the tactics of the opposing Jutland fleets to an appreciable extent. The use of oil in place of coal, for example, gave vessels a longer cruising range and, if need be, they could be refuelled at sea. In addition, greater flexibility was afforded to the control of steam pressure allowing for rapid increases of speed which was particularly useful in anti-submarine warfare. However, the greatest tactical advantage of oil-firing was the elimination of smoke through oil temperature and air supply during combustion – no longer could a fleet be given away by the tell-tale clouds of black smoke on the horizon. Although at Jutland there were generally sufficient coal-burning ships on both sides, such as HMS *Black Prince* and SMS *Von der Tann*, to nullify any degree of surprise, individual oil-fired vessels operating alone could get closer to the enemy than had previously been possible. This strategy continued right through to the Second World War when the use of air power and radar gave little opportunity for tactical surprise.

While the First World War did see the first use of air power at sea, it was effective mainly in patrol and reconnaissance roles rather than in direct combat. The Royal Naval Air Service made use of vessels like Seaplane Lighter *H21* (built 1918) to support seaborne aircraft operations. *H21* now forms part of the National Historic Fleet and is stored at the Fleet Air Arm Museum, Yeovilton.

Figure 12
The Spithead Review of 1924, at which George V presented colours to the Royal Navy. The Royal Yacht (HMY *Victoria and Albert*) leads the review party. Some of the 196 warships present had fought at the Battle of Jutland (1916), the largest fleet action of the First World War.

Figure 13
RMS *Queen Mary* at Long Beach, California. Built on the Cyde and launched in 1934, she crossed the Atlantic 1,000 times before being converted to hotel use and permanently moored. The *Queen Mary* is now included on the U.S. Register of Historic Places.
© David Jones, used with permission

Experimentation with the launching of aircraft from ships began before 1914, and in 1918 HMS *Argus* was launched, the world's first flat-top vessel capable of carrying military aircraft. By the Second World War the leading navies operated a variety of ever-larger aircraft carrier types.

Following German defeat in the First World War, 74 ships of the High Seas Fleet were interned in Scapa Flow, Orkney Islands. In June 1919, the fleet was scuttled on the orders of the German officer in command at Scapa Flow in order to prevent the ships falling into British hands following peace negotiations. Salvage of the ships commenced in 1922 but was halted by the Second World War. Some of the salvaged ships were lost after being raised, either because they broke tow or were abandoned, and are now represented in the wreck record. In 2001, the seven main wrecks were scheduled by Historic Scotland.

Following the end of the war in 1918 there was a need to replace some 7.7 million tons of British mercantile shipping which had been lost, and this provided work for many yards during the immediate post-war period. The number and size of new warships built was limited by the 1922 Washington Treaty. Although the evolution of ship and boat design slowed, there was a general desire to raise efficiencies of boilers and engines through the adoption of higher steam pressures and the use of superheated steam wherever possible. Improvements were made in boiler drum construction and in the design on combustion chambers, while innovation in the design of auxiliary machinery enabled the introduction of electric drive for main propulsion.

The first large British turbo-electric vessel was P&O's RMS *Viceroy of India* which entered service in 1928.

However, a general decline in world trade following the end of the war resulted in a reduced demand for shipping services, overcapacity and a consequent significant reduction in the number of new builds, compounded by the Wall Street Crash in 1929. The depression that followed saw consolidation and the closure of a number of shipbuilding yards across the UK. In 1921, for example, there were 15 yards on the Wear, but by 1937 this had decreased to only six. Although world trade began to increase towards the end of 1933, it was not until rearmament prior to the Second World War that the industry became fully revived.

1.3 Mid Twentieth Century (1933-1950)

Following the success of P&O's RMS *Viceroy of India*, Cunard responded with the RMS *Queen Mary* which captured the Blue Riband in 1938 by completing a transatlantic crossing with an average speed of 31.7 knots. A sister ship, the RMS *Queen Elizabeth*, was laid down in 1936 and, like the *Queen Mary*, spent the Second World War as a troopship. Both vessels were to give 20 years' service after the war. The *Queen Mary* is now permanently berthed as a hotel at Long Beach, California (Figure 13) while the *Queen Elizabeth* caught fire in 1972 and was subsequently partially scrapped. These were remarkable prestige vessels, but with the onset of feasable and much faster air transport, their domination of the market was under severe challenge.

The economies of operating a passenger or cargo service dictated the choice of steam engine to shipowners; passenger vessels were largely fitted with turbine engines for speed, while the higher effeciency of triple-expansion reciprocating engines were favoured by general cargo vessels right up to the end of the Second World War.

Another type of marine propulsion unit that gradually came into prominence during the inter-war period was the heavy-oil engine or diesel engine, named after Dr Rudolf Diesel who perfected it. Although first adpoted for marine propulsion in 1902, the conversion to diesel was a process that matured rapidly during and after the First World War, and in many countries, the diesel engine found increasing favour with ship owners in the 1920s and 1930s. Not only did diesel engines enable faster speeds, they were more economic and practical to operate than coal- or oil-fired (steam) ships.

The advantage of diesel engines was to prove decisive in the second half of the 20th century. The reason for this was that until the 1950s, state of the art marine engines could only run on high-grade diesel oil. This oil was unable to compete on costs with low-grade 'residual' oil burnt by steamships. The introduction of diesel marine engines designed to run on residual oil in the 1950s destroyed any cost advantage that steam might have had.

However, the relative merits of steam and diesel power assumed lesser importance after 1939 as the navies of the contesting powers had firmly adopted the steam turbine as propulsion units, such as that in HMS *Belfast*, a cruiser commissioned in 1939 that now forms part of the National Historic Fleet; the largest surviving Second World War warship in Europe, HMS *Belfast*, demonstrates how sophisticated such vessels had become.

Nine months into the Second World War, the Allied forces were stranded in northern France. Tasked with getting them home under Operation Dynamo, the Royal Navy requisitioned around 700 small craft to aid the larger warships in evacuating British and French troops from the beaches of Dunkirk. Several of the historic 'Little Ships', such as the *New Britannic* (which was built to carry 117 people but is credited with lifting, in several runs, some 3,000 men off the beaches) participated in the 2012 Thames Diamond Jubilee Pageant.

Throughout the war, the Allied need to build merchant ships at a rate faster than they could be sunk led to the adoption of standard power units such as the triple-expansion engines fitted in to the Liberty ships. These vessels were built in the USA and Canada in large numbers and include the designated dangerous

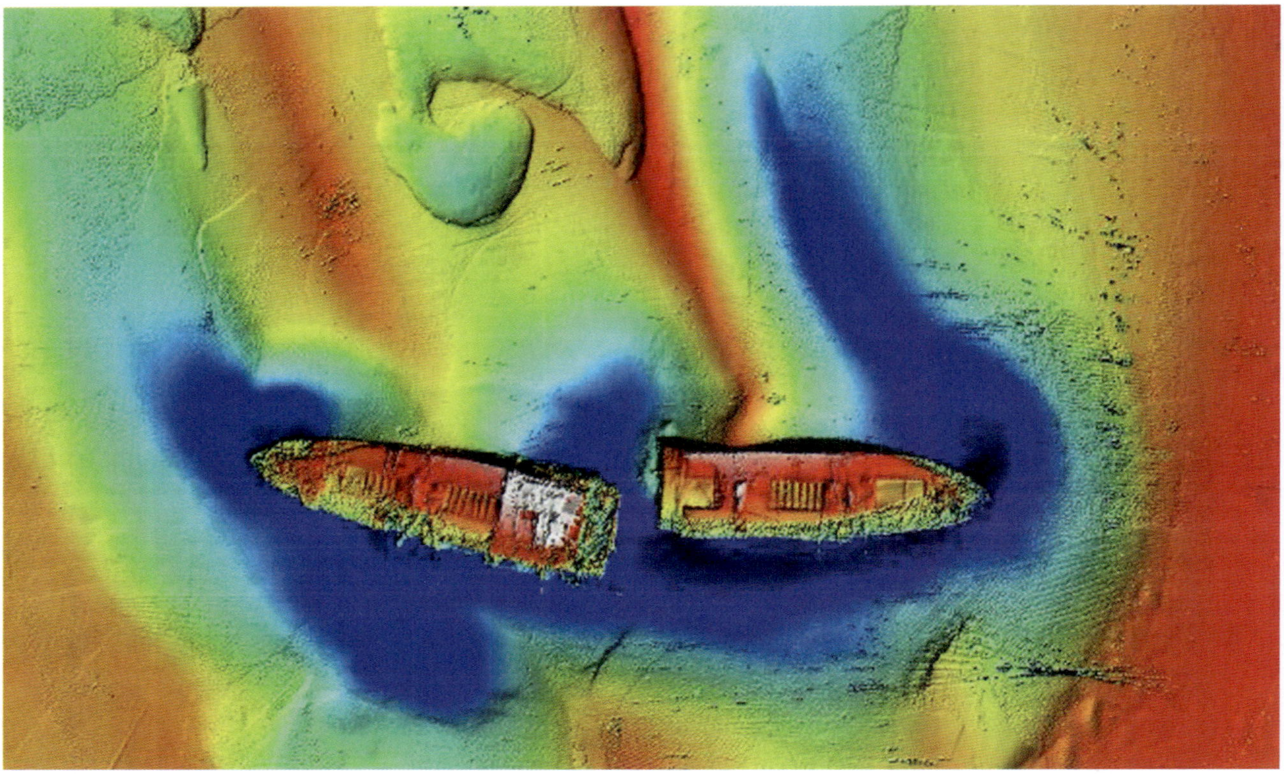

Figure 14
Multi-beam sonar image of the Liberty ship SS Richard Montgomery, in the Thames Estuary. On account of the presence of live ordnance the ship was designated a dangerous wreck in 1973.
© Maritime and Coastguard Agency

wreck SS *Richard Montgomery* in the Thames Estuary (Figure 14); this still carries much of its cargo of explosives. Advanced fabrication techniques, including a greater use of welding, reduced building time from years to months and, even though the service life of many of these vessels was brief, their contribution to the final victory of the Battle of the Atlantic was immeasurable. Over 2,700 were built.

The Admiralty concentrated on reliable standardised machinery layouts for all wartime construction, thus simplifying training and maintenance. British destroyer design during the war years was based on that of the successful J class vessels built in 1939 (none of which survive), and smaller Hunt class vessels were constructed specifically for convoy escort duties.

The extent and range of craft available to the Allies, including requisitioned vessels, can be most readily seen in the planning for Operation Neptune (the Naval component of Overlord). In December 1943, Admiral Ramsay submitted to the Admiralty an initial requirement for 467 warships plus 150 vessels for minesweeping duties. In fact, 702 warships (comprising battleships, monitors, cruisers, destroyers, sloops, frigates, corvettes, patrol craft and motor launches but excluding 25 flotillas of minesweepers) actually participated in the Operation. Representative of the Normandy landings, the remains of some Mulberry Harbour Units lie in areas of the south coast along the River Test at Dibden (Hampshire) and in Langstone Harbour (Hampshire), and include two Grade-II Listed Phoenix Units (reinforced concrete caissons assembled as part of the follow-up to the 1944 landings) in Portland Harbour and a scheduled Phoenix Unit at Shoebury Ness. Landing craft are represented in the archaeological record.

Fishing was restricted during the war but did not stop: the Hartlepool-based trawler *Isabella Fowlie* is recorded to have operated on a 'fishing and return' trip when bombed off St Abb's Head in 1941. The fishing fleets operating in the waters around England became increasingly dependent upon machinery in the 20th century, with fewer but bigger vessels. There were two broad types of fishing that took place during the mid 20th century. Firstly, local inshore fisheries comprising family-run businesses using oar and sail boats such as the *Boadicea* and, secondly, larger offshore fisheries operating steam-driven, and increasingly mechanised, vessels. However, many of the fishing vessels identified as wreck sites were actually requisitioned by the Admiralty for service during the two World Wars, principally as minesweepers and escorts; fishing vessels that were too small to be requisitioned or were otherwise unsuitable are likely to be relatively poorly represented archaeologically as wrecks.

During the latter stages of the war, the need for standardised construction to replace warship losses by enemy action was not so acute; designers were able to propose changes to machinery and layouts which took account of experience gained from the U-boat war. As a result, the *Weapon*-class destroyers which came into service in 1945 included many innovative features, such as two engine rooms, which were incorporated into other British post-war designs. The last of these destroyers, HMS *Crossbow*, was broken up in 1972.

In recognition of the loss of life associated with engagements during the war, many named surface vessels and submarines that were in military service at their time of loss – 78 in 2012 – are designated under the Protection of Military Remains Act 1986, while London's Tower Hill Memorial commemorates men and women of the Merchant Navy and Fishing Fleets who died in both World Wars and who have no known grave (Figure 15).

In the post-war years naval policy changed and the Admiralty settled firmly for frigates equipped with anti-submarine and anti-aircraft armament and with somewhat less power. Advances in metallurgy saw improvements in machinery components. Research devoted to the aero-engine realised the potential for combined steam and gas turbine machinery by the Admiralty which was introduced in a number of vessels laid down during the 1950s. Such an innovation had been proposed as early as 1912 but was halted because of the lack of satisfactory diesel units.

However, the monopoly of ships and boats as cargo-carriers was soon to be eclipsed. The conversion of ex-military airframes to transport people and cargo provided for long commercial flights as competion from the Douglas DC-3 and, from 1952, the de Havilland Comet saw the post-war re-introduction of scheduled services. From the early 1950s, transport by sea – particularly of passengers – could no longer hold up for rapidity against the international airlines. That said, container vessels – fast to load and unload – were introduced in the 1950s and now rival crude oil tankers and bulk carriers as the

Figure 15
The Tower Hill Memorial, Trinity Square, London, commemorates 35,749 men and women of the Merchant Navy and Fishing Fleets who died in both World Wars and who have no known grave. The First World War memorial (listed Grade II) was designed by Edwin Lutyens, and the Second World War memorial (listed Grade II*) by Edward Maufe.

largest commercial vessels on the ocean. Both developments impacted on port infrastructure, and continue to do so, in some instances rendering long-established ports like London no longer suitable for modern commercial use.

1.4 Submarines (1774-1950)

Important to navies for both offensive and defensive operations (as well as being used for rescue, research and tourism purposes) submarines offer the promise of invisibility, attack and escape. Records of attempted submergence date from 1578; the first, experimental, submarines were built in the 1620s, and fourteen types of submersible craft are known to have been patented in England by 1727.

By 1878, perhaps inspired by *Twenty Thousand Leagues Under the Sea*, published eight years earlier, a Manchester curate, the Rev. George Garrett, had set up a company to build a hand-powered submarine, the *Resurgam*. Garrett had intended to demonstrate the 12m long vessel to the Navy at Portsmouth, but the craft swamped and sank while under tow off North Wales. The *Resurgam* was discovered in 1995 and was designated a Protected Wreck Site the following year.

Many other submarines were built during the late 19th century by various inventors, but it was not to become a fully effective weapon until the 20th century following the invention of the internal combustion engine coupled with that of the electric motor, an effective storage battery and the Whitehead locomotive torpedo.

A key date in submarine history is 1864, when the Confederate submarine *Hunley* achieved the first successful sinking of a surface vessel. The first recognisably modern submarine dates from 1888 when J. P. Holland entered two US Navy competitions to design a submarine torpedo boat. He went on to form the Holland Torpedo Company and produced and tested several designs of Holland Class boats for the US Navy, leading to the *Holland 7* which was to form the basis of the Royal Navy's first five submarine boats (*Holland 1-5*).

Initially, the Royal Navy considered the submarine as 'an underhand form of warfare…and a damned un-English weapon.' The First Lord of the Admiralty, George Goschen, even commented that 'submarines are a weapon for maritime powers on the defensive.' This attitude was quick to change after submarines had entered service with foreign navies (including France). In April 1901, the new First Lord, Viscount Selbourne, announced the purchase of five Holland boats 'to assist the Admiralty in assessing their true value.'

The First British submarine, *Holland 1*, was launched on 2 October 1901, but the first to be commissioned was the *Holland 5*, which was also the first submarine in the Royal Navy to be fitted with a periscope. The Holland class boats were primarily used by the Royal Navy as a test-bed for early British submarine design and, like the following *A*, *B* and *C* classes, were fitted with a petrol engine (for surface propulsion) and an electric motor (for submerged propulsion).

The Holland boats served their purpose well, and even before the last of the type was launched the improved class that was to supersede them was already being built. Once their function had been fulfilled, the Navy quickly disposed of the entire Holland class; Nos. 1 to 3 were sold to shipbreakers, No. 4 had foundered in 1912 but was raised and expended as a gunnery target and the No. 5 foundered on 8 August 1912 while under tow to the Royal Naval base at Sheerness, Kent. Recovered in 1982, the *Holland 1* is now on display at the Royal Navy Submarine Museum, Gosport, and is part of the National Historic Fleet while *Holland 5*, the Navy's first commissioned submarine, was discovered in 2000 and designated as a Protected Wreck Site in 2005.

The Royal Navy's first British-designed submarine was the *Holland 6* (which was later redesignated *A1*). Launched in 1902, the *A1* was notable for the addition of a conning tower. Thirteen A-class submarines were built by Vickers between 1902 and 1905, and almost all were plagued by accidents and failures. The *A1* sank twice: first in 1904 when she became the first submarine casualty, with the loss of all hands; however, she was recovered but sank again in 1911, this time when it was unmanned. The wreck was discovered in 1989 and is now designated as a Protected Wreck Site. In January 1914, the *A7* sank with the loss of her crew during torpedo attack training. In 2001, the *A7* was designated a Controlled Site under the Protection of Military Remains Act 1986.

The *B*- and *C*-classes, launched between 1904 and 1910, were the last classes of petrol engined submarinies of the Royal Navy, and their construction marked the end of the development of the Holland-classes. By late 1916, the *B*-class submarines were no longer deemed suitable for combat and, with the exception of *B2* which sunk following a collision in 1911 off Dover, were sold for scrap. The only *C*-class submarine known in English waters is the *C11* which sank following a collision with the collier *Eddystone* off Norfolk in 1909. The surviving boats of the *C*-class were disposed of at the end of the war with the exception of *C4*, which was retained for trials until being scrapped in 1922.

German designers seemingly preferred to wait for improvements to kerosene, and later diesel, engines before developing submersible craft, as such engines are considerably more economical in fuel-consumption (meaning a greater range is given for a fixed quantity of fuel) and, more importantly, kerosene and diesel reduced to a minimum the possibility of the build-up of a flammable fuel/air vapor in a submarine. Germany's *Unterseeboote* U-1 was fitted with the first kerosene engine in 1906 but the diesel engine had become the standard propulsion unit of the Imperial German Navy's submarine fleet by 1910.

The Royal Navy's first diesel-powered submarines followed in 1907. These *D*-class submarines were also the first to be fitted with wireless transmitters and were capable of operating beyond coastal waters in the overseas patrol task which was of great importance in the defence of the UK's imperial and trade interests. Four were lost during the war, beyond England's waters, and the remainder were paid off in July 1919.

Steady improvements in design saw the British *E*-class submarines launched in three groups between 1912 and 1916, and these comprised the backbone of the Navy's fleet during the First World War. At the outbreak of war in 1914, Britain had 74 submarines, initially divided in to three groups to fulfill specific roles: overseas patrols, surface patrol flotillas working from the principal ports and harbour defence flotillas. As the war progressed, emphasis was placed almost entirely on the first role.

To combat the threat of the German High Seas fleet, a group of submarines was developed in 1913 to operate with the Grand Fleet having steam turbines for surface propulsion. These K-class boats were to suffer numerous fatal accidents and their overall contribution to the war was virtually nil. They were also the last submarines to be driven by steam until the advent of the Walter submarine developed in Germany during the latter stages of the Second World War.

Of seven known losses of British First World War submarines, the remains of only three have been located in English waters: *G3* (ran aground in Filey Bay, 1921), *G11* (ran aground near Howick, Northumberland, 1918) and *J6* (accidentally sunk by the Q-ship *Cymric*, 1918). In contrast, the wrecks of 41 German U-boats that sank during the war are known in English waters, comprising both coastal and mine-laying types. Two of these U-boats are designated under the Protection of Military Remains Act 1986: *UB-65* (sunk off Padstow, 1918) and *UB-81* (mined off the Isle of Wight, 1917). Other U-boats, such as U-*118*, were surrendered to the Allies at the end of hostilities and broken up (Figure 16).

Following the end of the war, many of the surviving British boats of the early classes were scrapped with attention being given to new designs beginning with the experimental *M* class, or 'Mutton Boats', as they were nicknamed. *M1* was fitted with a 12-inch gun which was intended for use against surface ships, *M2* was converted into a submersible seaplane carrier while *M3* became an exprimental minelayer. Following their accidental loss in the inter-war period during exercises off

Figure 16
Surrendered at the end of the First World War, *U-118* ran aground at Hastings in January 1919 when a towing cable snapped. It was broken-up where it lay in late 1919; some of the keel may still lie under the sandy beach.
© Hastings Chronicle, courtesy of Steve Peak

England's south coast, both the *M1* and *M2* have since become designated under the Protection of Military Remains Act 1986. The *M3* was scrapped in 1932. The only other British inter-war submarine losses in English waters were the *H*-class (the first to carry a bow salvo of four torpedo tubes) *Stock Force*, celebrated in the film 'Q-Ships' in 1928 in which the obsolete submarine was expended by gunfire off the Eddystone Lighthouse), and *L24* (sunk 1924 following a collision with the battleship *Resolution* off Portland).

Following further experimentation, such as the fitting of ASDIC (the acronym supposedly recalling the possibly fictitious Antisubmarine Detection Investigation Committee) underwater detection devices and VLF radio into the 1920s *Odin* class, British submarines became standardised into two main types: a medium size patrol type for operations in the North Sea, and a larger type to replace overseas classes.

At 11:00 am on 3 September 1939, all British submarines at sea received the signal: 'Commence hostilities with Germany forthwith'; there was unrestricted submarine warfare from the outset (whereas in the First World War that had only been waged from 1917). As in the First World War, the most effective and dangerous part of the German Navy was its U-boat arm, despite the Anglo-German naval treaty of 1935 limiting the *Kriegsmarine* submarine fleet to an equivalent 45 per cent of the Royal Navy's. The U-boat campaign was eventually out-fought by Allied naval and air forces, and out-built by Allied shipyards.

For recorded British submarines within English waters, losses during both World Wars were primarily accidental (for instance, due to navigational error, collision or non-hostile fire) while German losses were principally due to Allied attack. Only four Allied submarine losses are known in English waters from the

Second World War: HMS *Swordfish* (mined while on patrol off the Isle of Wight, 1940), HMS ***Umpire*** (sunk following collision off Norfolk, 1941), HMS *Unity* (sunk following collision off Wansbeck, Northumberland, 1940) and the French *Minerve* (ran aground off Chesil Beach, 1940). The *Swordfish* and *Umpire* are designated under the Protection of Military Remains Act 1986. Eighteen identified German U-boats lost during the Second World War are recorded in English waters (of which two are designated as military maritime graves under the 1986 Act: *U-1018* and *U-1063*), including the experimental *U-480* (mined in early 1945 off the Isle of Wight) which was equipped with a special rubber coating that made ASDIC detection difficult.

The end of hostilities brought a situation similar to that in 1918 where it was necessary for the British to scrap the majority of the submarine building programme. Primary interest lay in captured German submarines and, for those not scuttled under Operation Deadlight (where U-boats in British posession were deliberately sunk some 120 miles north-west of Ireland), extensive trials were carried out to test their performance with special attention being given to the Type XXVI Walter Boat which was propelled by a hydrogen peroxide-driven turbine.

While post-war development in the United States was to concentrate on atomic propulsion, the British continued to improve conventional types of submarine. HMS *Alliance*, commissioned in 1947, is now on display at Gosport Submarine Museum as a memorial to those who fought in similar boats and is also part of the National Historic Fleet. The later *Porpoise* and *Oberon* Classes were approved for building from the mid 1950s onwards and remained in commission for the Royal Navy until 1988 (*Porpoise*) and 1993 (*Oberon*).

1.5 Into the Nuclear Age (1948 and beyond)

The development of nuclear-powered craft – first submarines and subsequently surface vessels – has been seen as the single greatest achievement in the advancement of marine engineering by the post-war United States Navy. The application of atomic energy for marine propulsion is one of the fundamental advances in the history of transport.

Following construction of an atomic engine in 1948 in the US, the first nuclear-powered submarine, the USS ***Nautilus***, undertook sea trials in January 1955 though it took a further eight years for the Royal Navy to receive its first nuclear-powered submarine: HMS ***Dreadnought***. While nuclear-powered attack submarines gradually replaced the Royal Naval diesel-powered fleet during the Cold War, nuclear-powered civil merchant ships have yet to develop beyond a few experimental types.

Almost as an echo of the industrial past two hundred years earlier, steam power still survives in navies with nuclear-powered vessels, with the heat from nuclear reactions being used to raise steam to run a turbine. However, even when decommissioned, these radioactive modern craft will pose challenges for future curators and conservators for, as Ian Friel has observed, such relics can never find their way safely into any museum.

2 Further Reading

Good introductions to the subject include P Brown, *Britain's Historic Ships* (2009); P Brown, *Historic Sail: Britain's Surviving Working Craft* (2013); D Griffiths, *Steam at Sea: Two Centuries of Steam-Powered Ships* (2001); I Friel, *Maritime History of Britain and Ireland* (2003); P Kemp, *The Oxford Companion to Ships and the Sea* (1998); J Mannering (ed.), *Inshore Craft: Traditional Working Vessels of the British Isles* (2008); E McKee, *Working Boats of Britain: Their Shape and Purpose* (1983); K T Rowland, *Steam at Sea: A History of Steam Navigation* (1970) and T Crump, *A Brief History Of The Age of Steam: The Power That Drove The Industrial Revolution* (2007). Most known wrecks around the British Isles are listed in R Larn and B Larn, *Shipwreck Index of the British Isles* (1995-2000). For fixtures and fittings see M McCarthy, *Ships' Fastenings: from Sewn Boat to Steamship* (2005).

In terms of period overviews, the following are some of the more recent and valuable studies.

2.1 Victorian (1837-1901)

P Padfield, *Rule Britannia: The Victorian and Edwardian Navy* (2002)

R Parkinson, *The Late Victorian Navy: The Pre-Dreadnought Era and the Origins of the First World War* (2008)

Wessex Archaeology, *Assessing Boats and Ships 1860-1914*, unpublished report for English Heritage (2011)

R Woodman, *A History of the British Merchant Navy Vol. Three: Masters Under God: Makers of Empire: 1816-1884* (2009)

2.2 Early Twentieth Century (1901-1932)

A Couper *et al* (eds.), *The Conway History of Seafaring in the Twentieth Century* (2000)

D K Brown, *The Grand Fleet: Warship Design and Development 1906-1922* (1999)

R Gardiner, *Sail's Last Century: The Merchant Sailing Ship 1830-1930* (1993)

Wessex Archaeology, *Assessing Boats and Ships 1914-1938*, unpublished report for English Heritage (2011)

2.3 Mid Twentieth Century (1933-1950)

P Dawson, *The Liner: Retrospective and Renaissance* (2005)

J M Maber, *The Ship: Channel Packets and Ocean Liners 1850-1970* (1980)

J Thompson, *The War at Sea: the Royal Navy in the Second World War* (2003)

Wessex Archaeology, *Assessing Boats and Ships 1939-1950*, unpublished report for English Heritage (2011)

2.4 Submarines

P Akermann, *Encyclopaedia of British Submarines 1901-1955* (2002)

A N Harrison, *The Development of HM Submarines from Holland No. 1 (1901) to Porpoise (1930)* (Ministry of Defence BR3043; 1979)

2.5 Websites

https://www.gov.uk/government/policies/conservation-of-historic-buildings-and-monuments/ Department for Culture, Media & Sport historic environment webpages. [Site accessed: July 2016]

http://nationalhistoricships.org.uk National Historic Ships UK. National Historic Ships UK. The website includes databases on Registered Historic Ships and the National Historic Fleet and information about the selection process. [Site accessed: July 2016]

http://www.historicwarships.btck.co.uk/ Historic Warships Volunteer Group, based at The Historic Dockyard Chatham. [Site accessed: July 2016]

http://www.worldofboats.org/ Eyemouth International Sailing Craft Association, containing the former Exeter Maritime Museum collection of Ethnic, European coastal, Day Sailing and other interesting craft. [Site accessed: July 2016]

http://thesaurus.english-heritage.org.uk/ Historic England Thesaurus (for Maritime Craft Types). [Site accessed: July 2016]

http://archaeologydataservice.ac.uk/archives/view/marinemcd_eh_2008/index.cfm Online project archives for selecting marine sites that are sufficiently archaeologically important to warrant special measures, particularly the Selection Guide: Boats & Ships in Archaeological Contexts. [Site accessed: July 2016]

www.pastscape.org.uk National Record of the Historic Environment (NRHE). [Site accessed: July 2016]

3 Acknowledgements

Front cover: © David Noton, used with permission of the SS Great Britain Trust

Figure 1: © clydesite.co.uk

Figure 2: © Alasdair Moore, Tresco Abbey Gardens Valhalla

Figure 3: © Eyemouth International Sailing Craft Association

Figure 4: © Historic England Photo Library

Figure 5: © Historic England Archives

Figure 6: © Wessex Archaeology for Historic England

Figure 7: © Historic England Archives

Figure 8: © From the collection of Björn Larsson.

Figure 9: © Chatham Historic Dockyard Trust

Figure 10: © Mark Dunkley

Figure 11: © Mauretania Bar & Lounge, Bristol

Figure 12: © Historic England. NMR Aerofilms Collection

Figure 13: © David Jones, used with permission

Figure 14: © Maritime and Coastguard Agency

Figure 15: © Commonwealth War Graves Commission

Figure 16: © Hastings Chronicle, courtesy of Steve Peak

Contact Historic England

East Midlands
2nd Floor, Windsor House
Cliftonville
Northampton NN1 5BE
Tel: 01604 735460
Email: eastmidlands@HistoricEngland.org.uk

East of England
Brooklands
24 Brooklands Avenue
Cambridge CB2 8BU
Tel: 01223 582749
Email: eastofengland@HistoricEngland.org.uk

Fort Cumberland
Fort Cumberland Road
Eastney
Portsmouth PO4 9LD
Tel: 023 9285 6704
Email: fort.cumberland@HistoricEngland.org.uk

London
1 Waterhouse Square
138-142 Holborn
London EC1N 2ST
Tel: 020 7973 3700
Email: london@HistoricEngland.org.uk

North East
Bessie Surtees House
41-44 Sandhill
Newcastle Upon Tyne
NE1 3JF
Tel: 0191 269 1255
Email: northeast@HistoricEngland.org.uk

North West
3rd Floor, Canada House
3 Chepstow Street
Manchester M1 5FW
Tel: 0161 242 1416
Email: northwest@HistoricEngland.org.uk

South East
Eastgate Court
195-205 High Street
Guildford GU1 3EH
Tel: 01483 252020
Email: southeast@HistoricEngland.org.uk

South West
29 Queen Square
Bristol BS1 4ND
Tel: 0117 975 1308
Email: southwest@HistoricEngland.org.uk

Swindon
The Engine House
Fire Fly Avenue
Swindon SN2 2EH
Tel: 01793 445050
Email: swindon@HistoricEngland.org.uk

West Midlands
The Axis
10 Holliday Street
Birmingham B1 1TG
Tel: 0121 625 6870
Email: westmidlands@HistoricEngland.org.uk

Yorkshire
37 Tanner Row
York YO1 6WP
Tel: 01904 601948
Email: yorkshire@HistoricEngland.org.uk

Historic England

We are the public body that looks after England's historic environment. We champion historic places, helping people understand, value and care for them.

Please contact
guidance@HistoricEngland.org.uk
with any questions about this document.

HistoricEngland.org.uk

If you would like this document in a different format, please contact our customer services department on:

Tel: 0370 333 0607
Fax: 01793 414926
Textphone: 0800 015 0174
Email: customers@HistoricEngland.org.uk

HEAG133
Publication date:September 2012 © English Heritage
Reissue date: July 2016 © Historic England
Design: Historic England

Printed in Dunstable, United Kingdom

91218511R00020